An Introduction To The

GEOLOGY OF DEATH VALLEY

Text & Photography by Michael Collier

DEATH VALLEY NATURAL HISTORY ASSOCIATION
DEATH VALLEY, CALIFORNIA 92328

Racetrack Playa.

FOREWORD

Perhaps nowhere else on earth is geology better observed and studied than in Death Valley. The Valley may seem like a place where time has stood still for hundreds of millions of years, but the geologic story here is complex and reveals a land torn by great physical changes.

An Introduction to the Geology of Death Valley by Michael Collier captures the drama of the processes that are constantly changing Death Valley's landscape. He creatively weaves an account of Death Valley's one-hundred-and-forty-year mining history into the one-and-a-half-billion-year geologic history of the Valley. Through these stories and his perceptive observations of a slowly evolving landscape, there emerges a fresh perspective on Death Valley.

Michael Collier describes flash floods which create broad, fan-shaped deposits of sand, gravel and boulders; volcanoes blasting forth great volumes of ash that once blanketed the surrounding mountains and valleys; warm, shallow seas that extended over thousands of square miles. He describes a region rent by great earthquakes that pushed up mountain ranges and sank valleys; where thick sheets of solid rock have been pushed sideways for miles or gently slid down sloping mountain fronts; where the earth's forces have folded, crushed, or moved great masses of rock. As he so well puts it, "The soil of Death Valley—though scorched and salt-ridden—has been very fertile for geologic thought. Within Death Valley, rocks are deceptively barren, tantalizing in their degree of exposure. But each new strike only led to one more lost ledge; nowadays each new geologic theory leads to more questions."

On behalf of the Death Valley Natural History Association, I wish to thank Michael Collier for accomplishing an exceptionally difficult task. He has taken a complex, and sometimes controversial, geology and transformed it into story—the essence of solid natural history writing. The Association is proud to play a part in this quality interpretation of Death Valley.

George D. Curtis
Chairman, Board of Directors
Death Valley Natural History Association
1990

Library of Congress Number 90-081612
Editorial: Timothy J. Priehs
Design: Christina Watkins
Illustration p.18-19: Peter Corrao
All other illustrations: Deborah Reade
Typography: Triad Associates
Lithography: Lorraine Press, Inc.
Salt Lake City, Utah

Cover:
Hills above Artist Drive.

Back Cover:
Ridgeline on Telescope Peak.

Contents

Introduction

Most visitors to Death Valley are a lot like the prospectors who combed this country a century ago—they don't know just what they're getting into, but they do know that it's something big. Peaks reach a mile, even two miles, above the valley floor. Salt pans shimmer between mountain ranges, broken only by their own mirages. Hillsides glow green, gold and red in the slanting afternoon sunlight. If ever there was a land dominated by the presence of bare rock, illuminated as if from within by that old promise of "mineral," this must be it. From the bushiest-bearded prospector to the most casual tourist, everybody gets the message: in Death Valley, rock is everything.

On the assumption that good medicine has to taste bad, a lot of folks have stumbled through the worst that Death Valley could throw at them—searing heat, days without water, endless salt flats, impassable mountains—looking for the wealth that the land seemed obviously to hold.

From the time that the first '49ers crossed Death Valley in their headlong rush to the California goldfields, myths of immeasurable wealth were born and bred by this poker-faced desert. An 1849 party from Georgia and Mississippi that fancied themselves the "Bugsmashers" found a fabulously rich silver ledge that would come to be known as the "Lost Gunsight Lode"; but the party was barely able to get through the valley with their lives, let alone map out their discovery. The location of the silver—probably somewhere around Tucki Mountain—was forgotten. A young German emigrant named John Goller found nuggets of placer gold as he struggled through the southern Panamints, also in 1849. Sixteen years later, Charles Breyfogle crawled into Las Vegas Springs, more dead than alive with his scalp split open by the sun, beaten by Indians, and babbling about his "Lost Ledge" whose location he couldn't remember.

Stories of the Lost Gunsight Lode, Lost Goller, and Breyfogle's Lost Ledge would entice prospectors for decades. Discoveries were indeed made, but none could ever measure up to the myths themselves. After spending many a thirsty mile in search of the Lost Goller in 1909, veteran prospector Milo Page observed, "A lost mine is never really found, for the gold ledge that is once lost must forever glitter in the minds of men."

The prospectors tools were his burro and his pick. His geological expertise was anchored in notions of "rocks that have color."

Who were these prospectors? A few were misfits bent on squandering another man's grubstake money. But the prospector, as John Ross Browne observed, could be "a poet—though generally not aware of the fact. Ragged and unshaven, he owns millions, yet seldom has two dimes to jingle in his pocket—for his wealth lies in the undeveloped wilds. He scorns work, but will endure any amount of hardship in his endless search for 'rich leads'."

The prospector's tools were his burro and pick. His geological expertise was anchored in notions of "rocks that have color." He had an intuitive sense of contacts and faults that brought one rock up against another. He certainly knew quartz when he saw it; often gold or silver were associated with the quartz. But otherwise, the world was pretty much made up of hard rocks and soft rocks—finer distinctions he usually left to the geological engineers. On rare occasions he might find a ledge obviously sprinkled with visible gold or silver; more often he had to wait for the assayer to pass judgment on the value of a sample he had carried in from the mountains.

Most prospectors who made a strike were quick to sell out to the miners and speculators, gamble away the windfall and then move on. Always their dreams lay over the next ridge. In a 1951 book entitled *Loafing Along Death Valley Trails,* William Caruthers recounted an exchange he had with prospector Shorty Harris. Shorty told Caruthers, "When I go out, every time my foot touches the ground, I think 'before the sun goes down I'll be worth $10,000,000'."

"But you don't get it," Caruthers reminded him.

Shorty Harris stared at him with a sort of 'you're-too-dumb' look. "Who in the hell wants $10,000,000? It's the game, man—the game."

To these men, Death Valley was pure heaven posing as hell. Everywhere the rocks lay uncovered, waiting for their picks. Strata once buried and horizontal, here had been upthrust and exposed, intruded, folded and torn. The land was so chaotic that *anything* could lay over the next ridge. A hundred years later, the rocks confront us with every bit as much confusion and promise. Our ideas of wealth have evolved; now geologists mine Death Valley for knowledge, for insight into the past stretching a billion years back. The prospector, with his grey beard and grouchy burro, has gone over that next ridge.

THE SETTING

THE DESERT SOUTHWEST

> The Valley of Death, Inyo County, California, is the loneliest, the hottest, the most deadly and dangerous spot in the United States. It is a pit of horrors—the haunt of all that is grim and ghoulish. Such animal and reptile life as infests this pest-hole is of ghastly shape, rancorous nature, and diabolically ugly. It breeds only noxious and venomous things. Its dead do not decompose, but are baked, blistered, and embalmed by the scorching heat through countless ages. It is surely the nearest to a little hell upon earth that the whole wicked world can produce.

This poor fellow, working anonymously for the *New York World* in 1894, must have had a bad day. The rattlesnakes, he notes, are deadly, the scorpions enormous, and the Gila monsters hideous. He goes on to pronounce that the humidity is zero in Death Valley, the sun always shines and rain never falls. He is not all that far off the mark, even if, as it turns out, there are no Gila monsters here.

It can be warm in Death Valley. As much as 134° Fahrenheit in the shade at Furnace Creek on July 10, 1913. If you could find shade. Maybe hotter at Badwater, 100 feet farther below sea level, a hundred feet closer to hell. Officially it gets no hotter in the Western Hemisphere. Only two degrees hotter recorded anywhere else in the world. It can be difficult to get good data; in 1891 the weather bureau sent an observer to take observations during the summer months. After 153 days, he walked out before his appointed time, stating that he "would take hell straight" before ever again returning to Death Valley.

Joe Ryan once told William Caruthers that, "Heat is always exaggerated. Of course it gets a little warm now and then. Hottest I ever saw was in August when I crossed the valley with Mike Lane. I was walking ahead when I heard Mike coughing. I looked around. Seemed to be choking and I went back. Mike held out his palm and in it was gold nugget and Mike was madder'n hell. 'My teeth melted,' Mike wailed."

Salt deposits at Badwater.

At least it doesn't rain much: a little more than an inch and three-quarters per year. Some years there is no rainfall at a given spot in the valley. *None.* The little rain that does reach the ground is subject to an evaporation rate of 150 inches per year: a twelve-foot-deep lake, given no other source of water, would be dry in twelve months. Rainfall increases dramatically, however, with altitude—about two inches per year for every 1000 feet above 5000 feet. Summer rainfall frequently arrives in the form of thunderstorms, brief but intense downpours that can choke canyons with walls of angry muddy water.

Death Valley has not always been so forbidding. By examining the nests or ''middens'' of wood rats, we have learned a great deal about climate changes over the past few thousand years. Wood rats, also known as pack rats of the genus *Neotoma,* have a penchant for gathering all manner of debris and incorporating it into their middens. Since the rat forages at most only a few hundred yards from home, his midden gives us an excellent sampling of vegetation in the neighborhood. Middens have been found at elevations that are currently hot and dry. The middens, however, hold twigs, leaves, and needles of trees which today grow only at higher elevations with cooler, moist conditions. Carbon-14 dating of the plant remains show that they are 13,000 to 19,000 years old. The climate must have been cooler and wetter then.

Death Valley's present-day aridity is a direct function of its being in the ''rain shadow'' of the Sierra Nevada, Inyo, Argus, and Panamint mountain ranges. Each range sucks a little more moisture out of any storm blowing east in from the Pacific. Death Valley lies at the southwest tip of a much larger, relatively dry area known as the Great Basin. It covers virtually all of Nevada along with portions of Utah, Oregon, California, and a tiny bit of Idaho. True to its name, this region is an enclosed basin—a quarter million square miles without a single river draining to the sea.

When you think of Nevada, in the middle of the Great Basin, what do you see? Flat deserts?

Think again. Nevada, the heart of the Great Basin, has 413 mountain ranges. They are aligned north-south, likened by one geologist to a fleet of ships at anchor, all swung into a wind from the north. The Panamint, Black, and Funeral/Grapevine mountains of Death Valley belong to this fleet. Telescope Peak, at the top of the

The Great Basin has no outlet to the sea. This region is defined hydrologically: what few streams flow here, all lead to enclosed basins like Death Valley.

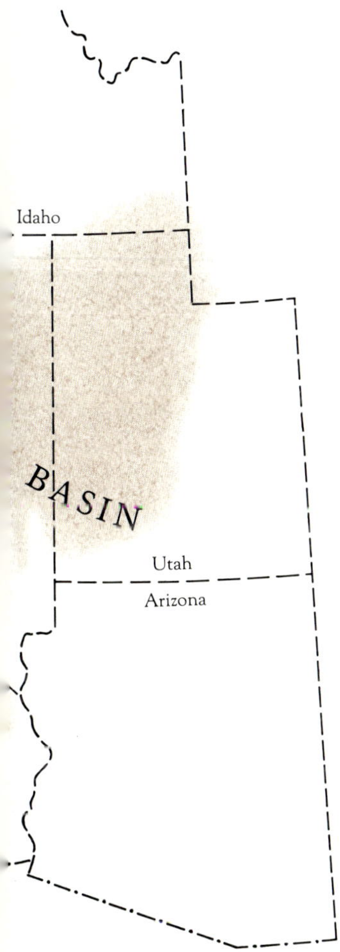

Panamints, is 11,049 feet above sea level. Wheeler Peak, at the top of the Snake Range in Great Basin National Park of eastern Nevada, is 13,063 feet high.

Of course Nevada and eastern California have their fair share of desert terrain. These deserts lie as basins between the ranges, their floors composed of the rocky debris shed from the flanks of the mountains. Death Valley and its neighboring Panamint and Amargosa valleys are just such intermontane basins.

Basin and Range. Great Basin. What's the difference? The Great Basin is defined *hydrologically:* none of its rivers run to the sea. The Basin and Range is defined *geologically:* a region full of parallel ranges separated by intervening basins; it just so happens that it includes much of the Great Basin.

The Basin and Range was defined as one of the country's dozen major geologic provinces more than a century ago, based solely on the casual similarity of its parallel mountain ranges. But as geologists further investigated its features, they realized that the province has a unique heritage. Its unifying characteristics go far beyond a few mountains that are coincidentally parallel.

The crust of the earth beneath the Basin and Range has been uplifted and stretched thin. It was extended perhaps as much as one hundred and fifty miles in the vicinity of Death Valley. The extension is directed west-northwest and east-southeast. The thinned crust allows heat to escape from the earth's interior at a remarkably high rate all over the province. The mountain ranges are all bounded by faults—great cracks in the earth's surface—that must somehow relate to this crustal uplift and thinning. Most of the mountain ranges have been tilted east or west by movement along the faults. In general, the mountains on the west side of the Great Basin are younger than on the east: the wave of faulting and uplift has steadily spread toward the west.

Geologists have been slowly piecing together a remarkable story of the evolution of western America. The plot calls for roiling subterranean caldrons, terrible earthquakes, and thundering volcanoes. Death Valley's rocks tell an important part of this story. But to hear them, we must first appreciate a few basic principles of geology. In many ways, these principles seem self-evident; but when applied to areas as complex as the Great Basin and Death Valley, they become

Death Valley may be the world's greatest wind tunnel. Wind not uncommonly howls up the valley at more than eighty miles an hour. The unbroken north-south alignment of the Argus, Panamint, Black, Funeral, Grapevine, Nopah, and Resting Springs ranges works to channel the winds along the basins that lie between the ranges.

Most sediments are laid down by running water, but sand dunes are deposited by wind. It's not an accident that sand dunes form only in a few particular areas of these desert basins. These dunes occur where the topography forces the wind to drop its load of airborne sand.

The sand dunes at Stovepipe Wells are a fine example of this phenomenon. Much of this sand is scooped up from streambeds in Panamint Valley and carried over Emigrant Pass into Death Valley. Winds blowing down Emigrant Canyon run into winds moving north through Death Valley. When these rivers of air converge, quiet eddies spin off in the lee of Tucki Mountain, and their load of sand drops out to form the Stovepipe dunes. The dunes here assume the crescent shape of a barchan dune, with their two points trailing downwind.

powerful tools in our attempt to sort out a convoluted geologic history.

If one rock is laid down upon another, then the higher one is necessarily the younger of the two. Likewise, if a rock is injected as a liquid that freezes in place within some other rock, the injected rock is necessarily the younger. Geologic processes that can be observed today are similar to those that operated in the past; these processes, like mountain-building and erosion, have been grinding away for tremendous spans of time, allowing slow processes to bring about great change. These simple ideas have been burdened with imposing names—the Principles of Original Horizontality, Superposition, Crosscutting Relations, and Uniformitarianism. Geologists hold to these principles with a zeal that borders on the religious.

Another fundamental observation is that there are three basic rock types—sedimentary, igneous, metamorphic. Death Valley has good representations of all three. Sedimentary rocks are made up of sand, silt, clay, lime, slime, and other materials that have been carried into place by water or air. The sediments are deposited in fairly flat layers, then compressed and cemented over time into rock. Igneous rock is created from molten material called magma, rising up from within the earth; it may slowly freeze underground and form large-crystal rocks like granite or diorite. Alternatively, magma might rise all the way to the surface as a volcano or lava flow, and cool quickly to form small-crystal rocks like basalt or rhyolite. Finally, metamorphic rocks are formed when an older rock is baked or squeezed, never so much as to melt the original rock, but enough that old grains are recrystallized into new minerals—and a metamorphic rock is born. Metamorphic rocks like schist are often finely striated by the parallel alignment of their minerals; this alignment is visible to the naked eye as a set of planes called foliation.

One further fundamental consideration: what's time to a geologist? Humans, according to various scientific and religious persuasions, have been around for perhaps 300,000 years. Our closer forebears are recognized through a bit more than 2 million years of the immediate past. Most geologists have an odd sense of time. To them, a couple million years is barely a drop in the bucket. Geologists divide time into four great blocks, all hinged on the appearance and characteristics of fossils contained in the world's rock record.

Fossils of hard-shelled animals appeared very suddenly in rocks

The Basin and Range is a geologic province that covers parts of seven states and northern Mexico. It is characterized by a repeated pattern of basins nestled between mountain ranges.

all of about the same age. Animals certainly existed earlier but, without hard shells, they were rarely preserved as fossils. All time before this sudden appearance is called the Precambrian. Time since has been divided into three eras—Paleozoic, with old life forms like trilobites and early fish; Mesozoic, with its dinosaurs and early mammals; and the Cenozoic, with recent life forms like camels, horses, and humankind. The Pleistocene is an important subdivision of the Cenozoic, comprising the relatively recent Ice Ages.

These divisions of time, based on fossils, are adequate for showing that one rock was older or younger than another. But calculating a specific age for each era required the development of another fundamental tool of geology: radiometric dating. Radioactive particles decay at a constant rate, not influenced by heat or pressure which change the rates of other processes. Geologists have been able to measure minute quantities of radioactive elements and their break-down products within rocks, and then calculate the age of the rock in years. No longer do they always have to rely on the earlier terms of 'older' or 'younger'. We can now assign numerical ages to the four geologic eras: Precambrian, before 570 million years; Paleozoic, 570 to 225 million years; Mesozoic, 225 to 65 million years; and Cenozoic, 65 million years to the present.

GEOMORPHOLOGY

THE LAY OF THE LAND

I rarely get dizzy. But the Racetrack Playa almost did me in. The playa—a dry lake bed—is about as flat as flat can get. I had walked midway across, trying to catch up with an odd school of rocks that is known to swim along the playa's surface, when I felt the first waves of disequilibrium wash over me. With nothing to break an absolutely flat foreground, no rocks or gullies or hillocks, my balance began to erode. I kept both eyes on the playa's edge a mile away. My head seemed to float just above the line stretching to the horizon.

There are many stories of parched men lost in the heat of Death Valley who stumble into some prospector's camp, holding their clothes above their head. With hallucinogenic fervor, they explain that they are immersed in water, floating just high enough to breathe, barely able to keep their clothes dry. Flat land—especially *hot* flat land—is not always conducive to good health. Wandering around the Racetrack, I could empathize with those naked lost souls. At least I still had my clothes on.

The Racetrack is an enclosed basin: water flows in but must evaporate out. Its surface is composed of very fine silt which, when pounded by a good soaking rain, becomes remarkably slippery. Water fans out, an inch or two deep, perfectly flat. Winds gusting down Hunter Mountain and out of Saline Valley can be strong enough to set rocks sailing in from the playa's edge, tacking first this way, then that. As the playa dries, the rock trails are frozen into its surface until the next storm arrives.

Death Valley itself, fifty miles to the southeast, is floored by an even larger playa. Like the Racetrack, Death Valley has no outlet to

The lowest parts of Death Valley are remarkably flat. Water flowing downhill must at some point reach its final destination—a level that geologists call base level. For most of the world this is sea level. But no rivers run from Death Valley to the sea; instead water establishes its own new base level, which happens to be 282 feet *below* sea level in Death Valley.

When flowing water reaches base level, it can no longer carry sediments. In Death Valley, these sediments fall out to form the flat bottom of a playa; the remaining water can only escape by evaporation. Minerals that had been dissolved in the water are necessarily left behind on the playa. These evaporites—salt, gypsum, trona, and others—form the fluffy white deposits that are so prominent on the floor of Death Valley.

This picture illustrates the meandering courses of water flowing toward Badwater from the Stovepipe Wells area. The meanders are a function of the playa's flatness: with little gradient, the streams just can't make up their minds. The brown areas have a higher concentration of silt and sand, whereas the white areas are more completely covered by evaporites.

the sea. Indeed its lowest point near Badwater is 282 feet below sea level. The Amargosa River, aided by ephemeral washes and a few permanent springs, carries sediments down from the surrounding mountains into the valley. The sediments are left behind as the water evaporates; they would have completely filled Death Valley long ago had not the floor itself been steadily sinking faster than the sediments accumulate.

Dante's View offers a wrap-around perspective of Death Valley, one mile below. From here you can see the stark white outline of the valley's saltpan, stretching from Salt Creek south almost to Shoreline Butte and covering some two hundred square miles. The course of the Amargosa River, flowing in from the south, blossoms into a lacy pattern of distributaries as it nears Badwater. If you look carefully, you may see the reflection of water in Salt Creek or along some stretches of the Amargosa.

Death Valley holds more water than might at first meet the eye. The Amargosa River, rising north of Beatty, Nevada, describes a great arc to the south before making up its mind to flow west and then north into Death Valley. All told, it drains some 6000 square miles, and appears to receive underground contributions from an even larger area. Its perennial stretches near the Confidence Hills might carry as much as a few cubic feet of water per second. Salt Creek drains into the northern end of Death Valley; it can run inches deep and still be a few feet wide. Don't laugh; after all this is a desert.

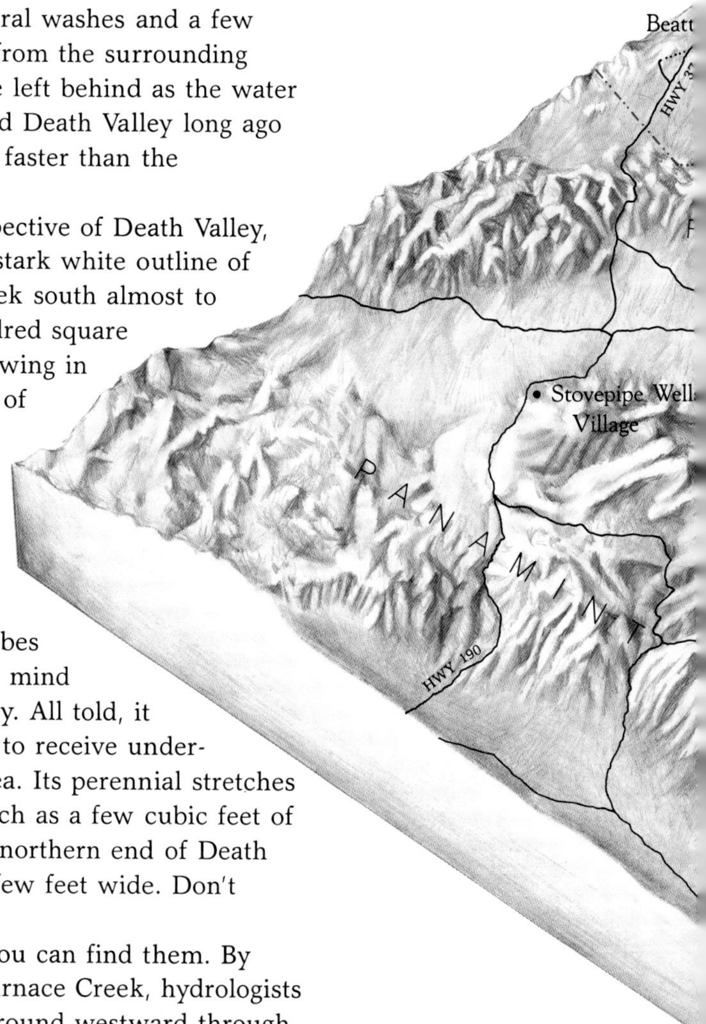

Numerous springs rim Death Valley—if you can find them. By examining trace elements in the springs at Furnace Creek, hydrologists have concluded that water is flowing underground westward through fractured zones in the Funeral and Black mountains, from the area around Ash Meadows towards Furnace Creek, rather than taking the longer course described by the Amargosa River.

As water flows into the basin, evaporation takes its toll, steadily concentrating the remaining brine. Minerals crystallize out of solution as they reach their individual solubilities. The basin's deposits are therefore roughly arranged in a set of concentric rings. Calcite

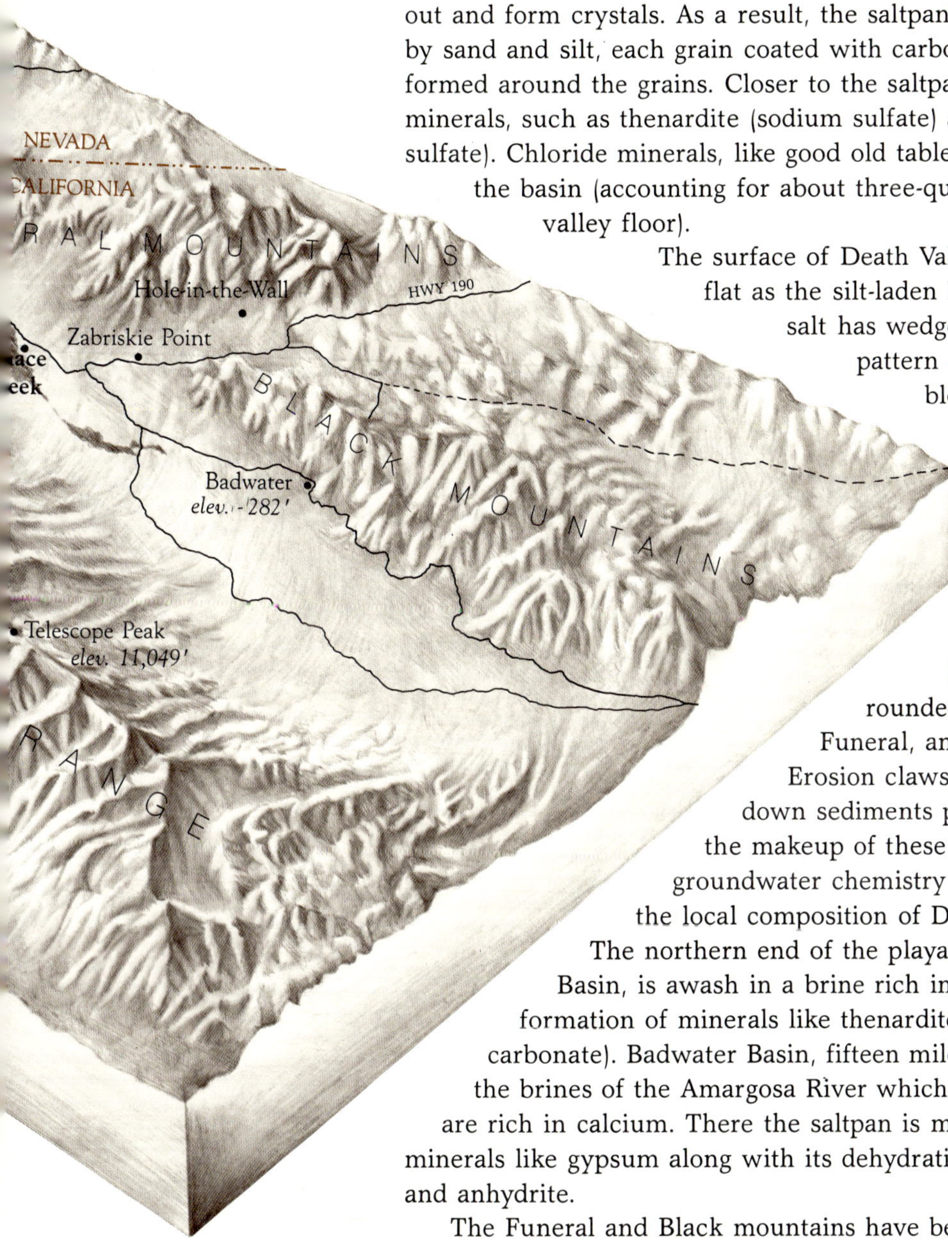

(calcium carbonate) is among the first of the playa's minerals to drop out and form crystals. As a result, the saltpan's outer rim is marked by sand and silt, each grain coated with carbonate minerals that have formed around the grains. Closer to the saltpan's center are sulfate minerals, such as thenardite (sodium sulfate) and gypsum (calcium sulfate). Chloride minerals, like good old table salt, cover the center of the basin (accounting for about three-quarters of the entire valley floor).

The surface of Death Valley's saltpan is not so flat as the silt-laden Racetrack. In places the salt has wedged itself into a mosaic pattern of polygons, like jostling blocks of flat arctic ice. At the Devil's Golf Course, the playa has been alternately flooded then dried, creating an impossibly rough surface of salt spires and castles.

Death Valley is surrounded by the Panamint, Funeral, and Black mountains. Erosion claws at their flanks, washing down sediments peculiar to each range; the makeup of these sediments influences groundwater chemistry and consequently alters the local composition of Death Valley's saltpan.

The northern end of the playa, called the Cottonball Basin, is awash in a brine rich in sodium, leading to the formation of minerals like thenardite and trona (a sodium carbonate). Badwater Basin, fifteen miles farther south, receives the brines of the Amargosa River which, though poor in sodium, are rich in calcium. There the saltpan is more likely to contain minerals like gypsum along with its dehydration products, bassanite and anhydrite.

The Funeral and Black mountains have been witness to a good deal of volcanic activity. Five or six million years ago, local volcanic rocks were eroded and redeposited within the Furnace Creek Formation which is now exposed in areas adjacent to Cottonball Basin. As a reflection of its volcanic ancestry, this formation contains a high concentration of boron. Ground water flowing across the Furnace Creek Formation dissolves the boron and then deposits it in the northern end of the Death Valley playa.

Borate minerals—like ulexite (a.k.a. 'cottonball') and borax—have been prized for at least the past four thousand years. Babylonians used borax as a flux in their welding of jewelry. The Chinese glazed pottery with it for a thousand years before Marco Polo brought borax crystals from Mongolia to Europe. More recently, boron compounds have been used in the manufacture of everything from soaps to gasoline, herbicides to pharmaceuticals, water softeners to pyrex glass.

Death Valley's name will always be linked with lost gold and silver mines, fantastic bonanza strikes by grizzled old prospectors. But the valley's greatest mining wealth lay in its lowly—and extensive—playa deposits of ulexite, borax, and kernite. Isadore Daunet discovered ulexite near Bennett's Well in 1875, but the value of his discovery at first went unheralded. In 1881 he got around to building the Eagle Borax Works, though he managed to recover borax from the ulexite for a mere two years. More successful were the Harmony Borax Works, discovered by Aaron and Rosie Winters in 1881, and put into production by William Coleman the following year.

Coleman's greatest obstacle was the distance—165 miles—between his borax works and the nearest railhead at Mojave. His superintendent J.W.S. Perry oversaw construction of wagons weighing 7800 pounds, capable of carrying as much as 46,000 pounds of cargo. When hitched to eighteen mules and two horses, these became the Twenty Mule Teams that operated out of the Harmony Borax Works from 1884 until 1888 without a single breakdown.

Ulexite mining was a primitive affair: deposits were simply scraped from the surface, dissolved in water, and then distilled and hauled off to market. In 1882 a new borate mineral was discovered and, surprisingly enough, named "colemanite." This contained higher concentrations of boron. Beginning in 1886, huge deposits were mined at Borate, California, south of Death Valley, leading to the closure of the less competitive Harmony Borax Works by 1888. The Twenty Mule Teams were put out to greener pastures.

Anatomy of a Basin.
The floor of Death Valley has been subsiding for the last few million years, sliding down the west face of the Black Mountains. As the bedrock floor drops, sediments shed from the surrounding mountains accumulate in the basin. Once deposited, these sediments are themselves tipped as the floor continues to rotate downward.

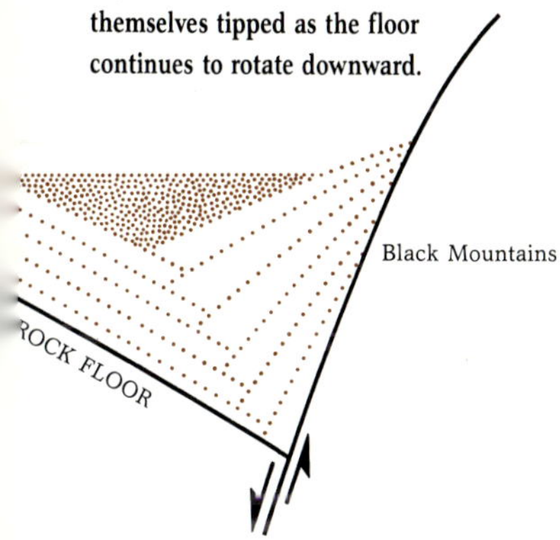

As the mines at Borate were depleted around the turn of the century, new claims were developed in the northern Black Mountains at the Monte Blanco and Lila C. mines. These were located, not as playa deposits on the easily accessible valley floor, but as underground vein deposits within the Furnace Creek Formation. They required the added effort of underground mining.

Abortive efforts to haul ore with a steam-powered tractor called "Old Dinah" led to formation of the Tidewater and Tonapah Railroad. This railroad never quite managed to reach either Tidewater or Tonapah, but did succeed in hauling a third of a million tons of colemanite from the Lila C. mine, worth $8,000,000—more than all of Death Valley's other mines combined.

With borax, Death Valley had achieved a spot in the mining limelight never afforded by the more precious metals. Gold and silver camps had come and gone a'glittering, each with its history of wheeling-and-dealing and sporadic violence. In contrast, Ryan, the company town serving the colemanite mines of upper Furnace Creek, was a model of peaceful living.

John Randolph Spears, writing in 1892, typified this bucolic way of life when he described a few of the early borate prospectors:

> What is known as the Monte Blanco deposit of borates in Furnace Creek Canon was located by Philander Lee, Harry Spiller, and Billy Yount. Tradition does not make them noted in any other way. They never killed anybody, never got killed, never got wounded even. They just lived, found a deposit of borates and sold it for $4000. Philander used his share in making a ranch at Resting Springs. The other two "went off somewheres." Philander is an interesting fellow, for he has a squaw and several half-breeds, and a brother named Leander...and two other brothers named Meander and Salamander.

Folks have probed and prodded the depths of Death Valley, attempting to define just how far down the sediments of the playa stretch. Bedrock—the feet of the surrounding mountains—must be down there somewhere. Drillers have chewed through a thousand feet of salt, clay, salt, mud, muck, salt, clay, and then more salt. All without reaching bedrock. Geophysicists tell us that dense rocks (like

Death Valley is justifiably famous for its classic alluvial fans. In their purest forms, these are the semicircular deposits of debris shed from the mountains that surround the valley. Though normally dry, once in a great while the canyons that drain the mountains are brimful with the floodwaters of intense local rainstorms. The floods roar through the confines of the canyons, and carry surprisingly large loads of sand, gravel, cobbles, and even boulders.

Within the canyons, the streambeds are steep and the water flows fast. But when the water reaches the canyon mouth, the gradient suddenly becomes shallow; the water spreads out and loses velocity. Consequently it dumps sediments it had previously been able to carry, forming an alluvial fan. Over time each successive flood is forced to one side or the other of the earlier deposits, and the fan gradually assumes its classic conical shape.

The sediments carried down the canyon are a hodgepodge of rocks of all sizes that have had little time to be smoothed and rounded by the action of running water. This angularity and poor sorting is a hallmark of deposits within an alluvial fan. Thus, when geologists examine an old rock layer and find these same characteristics, they can conclude that they were deposited as an alluvial fan.

bedrock) have more gravitational pull than light rocks (like Death Valley's loose sediment and salts). Using sensitive gravimeters, they have shown that the valley's bedrock floor is tilted down to the east. Because of this eastward tilt, the sedimentary fill is deepest beneath Badwater. Here the sediments are 9000 feet thick.

Nine thousand feet. Telescope Peak, only fifteen miles to the west, today stands more than two miles above the valley floor. If the sediments that fill Death Valley were scooped out, and you could stand on the exposed bedrock, you would be looking 20,000 feet up to Telescope Peak. But you might as well forget it; the National Park Service would never let you get away with it.

From the drillers' record of salt and mud, a fascinating history of Death Valley's playa has been pieced together. The Pleistocene Epoch, about 2½ million to 10 thousand years ago, was marked by a world-wide shift toward a colder climate. The Sierra Nevada Mountains experienced heavier precipitation and significant glacial advances. Meltwaters poured down the mountains' east side, overfilling Owens Valley, China and Searles lakes, then Panamint Valley. When Panamint Valley's Pleistocene lake was 900 feet deep, it would in turn overflow into Death Valley around the southern end of the Panamint Mountains. A lake, posthumously named Lake Manly, stood as much as 600 feet deep in Death Valley.

Death Valley and the surrounding basins are host to a curious genus of desert pupfish, found isolated within springs like Saratoga or closed drainages like the Amargosa River. The fish evolved from a single line of ancestors; each spring or river now has its own species of pupfish, suggesting that these bodies of water have been separated long enough for evolution to have worked its magic.

Lake Manly's legacy is written on the walls of Death Valley. Shoreline Butte, west of the Ashford Mill site, is striped with many thin horizontal terraces, each eroded by the lake's waves when its surface stood at a given elevation. The Black Mountains above Badwater and Mormon Point are similarly incised with terraces. Low hummocks were formed by currents within the lake running near its shores; these gravel bars can be seen on the road to Beatty, two miles north of its turn-off from Highway 190.

The lake (or *lakes*) came and went, depending on vagaries of the Pleistocene climate and rates of evaporation. Separate high stands of

Lake Manly have been tied to especially moist climate phases, roughly 20,000 and 130,000 years ago. Black lake-bottom muck was laid down on the valley's floor. This muck is now seen as a twenty-foot-thick section in the drillers' cores, beneath the salt deposits that have accumulated since Lake Manly disappeared 10,000 years ago. The playa has been covered only by ephemeral lakes since Pleistocene time, usually inches deep and lasting but a few weeks after major storms. One lake, 2000 years ago, may have lasted a bit longer since it was thirty feet deep.

Death Valley is a naked land. Geologic processes, elsewhere occurring under cover of soil and vegetation, here proceed in full view. Most visitors are deeply impressed by Death Valley's dearth of moisture. But sudden storms can erase that impression in a flash. The Twenty Mule Team skinners talked of twenty-foot walls of water crashing down Wingate Wash. Titus Canyon Drive, its walls in one place only eighteen feet apart, is necessarily closed during thunderstorm season every year, because of the very real risk of flash floods filling the canyon.

Flash floods are the workhorse of desert erosion. These floods are briefly able to carry tremendous loads of debris, ranging from sand grains to house-size boulders. But as the flood waters burst beyond the confines of their mountain canyons, they fan out toward the valley floor; the water loses velocity and drops its load of sediments. This accumulation of sand, cobbles, and boulders gradually assumes the conical shape of an alluvial fan.

The fans' scratched and scarred surfaces show multiple episodes of deposition followed by renewed cutting and stream reentrenchment back down into the fan. The oldest fan surfaces tend to be left as remnants high up against the mountains; the fans are usually youngest near their toes at the valley floor. The older, higher surfaces have been darkened by "desert varnish," a manganese and iron oxide deposit concentrated by microbes and fixed to the rock surface by clay minerals.

These fans are most extensive on the west side of Death Valley, stretching in great aprons down from the eastern side of the Panamint Mountains. The alluvial fans on the other side of the valley, at the foot of the Black Mountains, tend to be much steeper, more abrupt and abbreviated. The differences are best explained by the tipping of the valley's bedrock floor. As will be discussed in more depth later, the rising of the Panamint and Black mountains has been accompanied by an eastward tilting of the mountain blocks.

Gravity studies have shown that the boundary between sedimentary fill and bedrock is deeper beneath Badwater and the eastern side of the valley, than on the west adjacent to the Panamint Mountains. This eastward tilting continued as the alluvial fans have grown out from each mountain range. As a result, the Black Mountain fans are steep, their bases buried deep in the basin, while the Panamint fans are shallow, gently sloping much farther out into the valley. This asymmetry is reflected in the position of Death Valley's saltpan, snugged up against the foot of the Black—not the Panamint—Mountains.

Death Valley owes its existence to the coupled movement of two faults: the Furnace Creek Fault Zone seen here beneath the Grapevine mountains, and the Southern Death Valley Fault Zone at the south end of the Monument. Both are strike-slip faults, with movement that sends rocks on the left side of the fault northbound, and rocks on the right side of the fault southbound.

Alluvial fans that spew down from the Grapevine Mountains (on the right) converge with one another into a pediment surface which has been re-dissected by stream channels wandering across the surface on their way down to the Death Valley playa. The toe of this pediment has been cleanly clipped by the fault.

STRUCTURE

BARE BONES OF THE EARTH

The wind began to pick up late in the afternoon. By sunset, I was curled up at the base of a bristlecone pine, staying clear of the full-on gale that howled through the branches above me. I spent the night at 11,000 feet, on the lee side of Telescope Peak. Badwater lay two miles directly beneath me. The view was staggering. I realized that Death Valley National Monument is more mountain than valley: there were the Black Mountains, Funeral and Grapevine mountains, Panamint Mountains. Throw in the Greenwater, Cottonwood, Hunter, and Last Chance mountains for good measure. The pyramid-shaped purple shadow of Telescope Peak slid across the Badwater Playa, up the west side of Smith Mountain. Looking out over this endless series of ranges stretching into the eastern dusk, I was alone and felt far from home. Night pulled itself like a cover over all that is Death Valley.

The story of Death Valley geology is, in large measure, the story of its mountains. Each range is unique: the rocks found within the Black Mountains are radically different than those found in the Panamints. The Funeral Mountains contain thousands of feet of Paleozoic dolomite, a magnesium-containing sedimentary rock; the Greenwater Mountains just a few miles to the south are covered by younger volcanic rocks. Looking down from Telescope Peak, I saw rocks that are a billion years old, set alongside other rocks which, to a geologist's mind, might as well have been formed yesterday—only a few hundred thousand years old.

Death Valley's prospectors were certainly aware of these sometimes subtle differences in the mountain ranges; these men succeeded or failed by the strength of their geologic intuition. It was the rare prospector who arrived in the field with formal geologic training. William Lewis Manly, one of the party that survived—and subsequently

We often think of faults as being colossal breaks in the earth's surface running for miles at a stretch. But in Death Valley it's possible to become acquainted with faults in a much more intimate way.

As rocks slide along a fault plane, one rock will commonly gouge the other. Friction caused by the sudden movements creates a fine striated and polished surface called a slickenside. With your eyes closed, you can run your hand across the surface and know which way the rocks once moved.

This twenty-foot cascade of slickensides adorns the wall of a side canyon in the Black Mountains. The Furnace Creek Formation here was faulted as the mountains were uplifted. This canyon eroded along a weak zone created by the fault. For the most part, the Furnace Creek Formation is hard enough to preserve such slickensides; in areas where a fault crosses softer rock, the slickensides are usually erased quickly by erosion.

named—Death Valley in 1849, was lured back eleven years later, looking for the Lost Goller mine. He commented that, "As I had been in the gold mines for several years, I could tell a gold-bearing country." All the same he didn't find the Lost Goller.

Shorty Harris was chasing his unruly burros above Old Man Beatty's ranch in 1904, when he stumbled onto a ledge that was lousy with gold. "It seemed to me that the whole mountain was gold. The formation had a good dip, too. It looked like a real fissure vein; the kind that goes deep and has lots of real stuff in it."

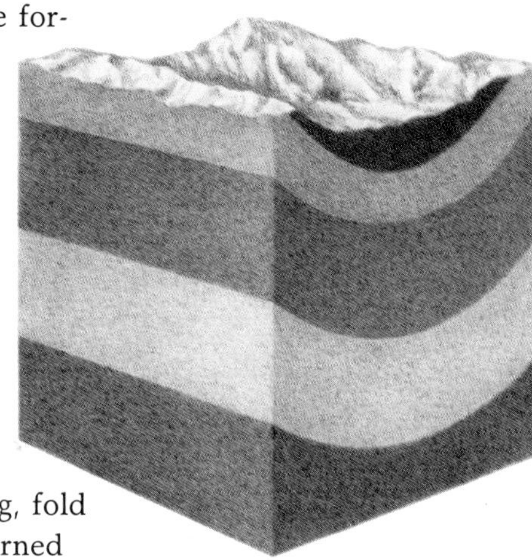

Formation. Dip. Vein. Gold. This was the language and lifeblood of prospecting. These fellows learned to look for breaks in the rock called faults, where rocks on one side of the fracture had ground past rocks on the other side. The fault itself might be gouged and broken, or perhaps polished into a smooth streaked surface called a slickenside. Gold could have been deposited along such a fault. The prospectors learned to watch for the white intrusions of quartz-rich magma that cut up into surrounding country rock, maybe bringing gold with it. Gold-bearing ledges could be exasperating, fold upwards to be eroded or dip down out of sight. These men learned their geology and earned their gold the hard way—by trial and error with a pick and shovel.

Rocks once deposited as flat layers can be squeezed and bent. A syncline (above) bows downward, while its mirror image, the anticline (right) bows upward. Walking across a level surface that cuts across an anticline, you would move from younger to older and then back to younger rocks.

Shorty's ledge, by the way, turned out to be the Bullfrog, around which surged a huge gold rush that spawned the town of Rhyolite, Nevada. A well-liquored Shorty, according to some reports, soon sold his interest in the claim for $700 and a burro.

Folded rock! Rocks, in our everyday experience, don't behave like taffy. But given sufficient time and pressure, even rocks will bend. And subjected to repeated episodes of pressure from different directions, these folds will in turn be folded and then re-folded. The tortured rocks of Tucki Mountain record seven phases of deformation, each phase further obscuring the previous one. Sorting out this kind of geology is much like looking at a dried-up piece of used chewing gum and proving that it once was flat. Hard work.

Driving east out of Furnace Creek, near the entrance to the Twenty Mule Team Canyon Drive, you will see rocks tipped steeply towards the north. Here beds of the Furnace Creek Formation dip sixty and seventy degrees into the earth, north side down. Three miles to the north at Hole in the Wall, the same formation also dips sixty or seventy degrees, but there dipping down to the *south.* The formation was originally deposited as a continuous, roughly horizontal sheet. Since deposition, it has been folded downward in the center, forming what is known as a syncline. The center of the syncline is now covered by younger detritus carried down by Furnace Creek Wash. But if you could drill deep enough beneath the wash, between the outcrops at Twenty Mule Team Canyon and Hole in the Rock, you would expect to find the same rock layer buried at the low point of the syncline.

The geometric complement of a syncline is called an anticline, bowing upwards instead of down. Levi Noble, a gifted geologist working in Death Valley in the 1930s and '40s, outlined the Desert Hound Anticline north of Jubilee Pass. Given the assumption that older rocks usually lie beneath younger rocks, we can differentiate a syncline from an anticline even more precisely: in a syncline, the center of a horizontal crosssection will hold younger rocks, while in an anticline, the center of such a crosssection will hold older rocks. This becomes important in the event that either of these folded structures might have been turned upside down.

Rocks, of course, are going to bend any which way they please. An overturned or recumbent fold is preserved at the very top of Corkscrew Peak, just west of Daylight Pass. Here the Principle of Superposition meets its most serious challenge. The youngest rocks are supposed to be on the top—right? The diligent geologist, carefully studying his fossils as he climbed, would find progressively younger fossils as he went higher *until* he crossed the axis of the fold. Any higher, and he would be climbing up into older fossils and rock. The folding at Corkscrew Peak has turned sections of the Carrara Formation completely upside down.

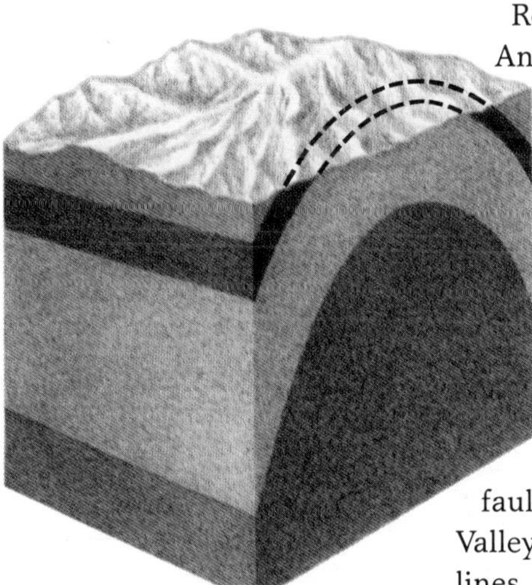

Death Valley is certainly not without its faults. Indeed, fault-finding may well be the geologist's favorite pastime in Death Valley. What does a fault look like? One trick is to watch for straight lines. Straight lines aren't often preserved along the earth's surface without an underlying reason. A canyon that cuts straight back into a hillside may be following rock weakened by a fault. A sudden step-down that slices across an alluvial fan (called a scarp) may be caused

A nameless stream flows out of the Funeral Mountains into Furnace Creek. Along its course down from the flanks of Schwaub Peak, the stream encounters the tipped strata of the Furnace Creek Formation. Over time the softer layers of this formation eroded more quickly than the resistant ridge, and the stream was forced to cut the gash that is the Hole in the Wall.

These layers of the Furnace Creek Formation were deposited four to six million years ago within a trough that was sinking even as the rocks were being deposited. The sediments were initially laid down in almost-horizontal layers. But as the trough deepened, the layers were progressively folded down to form what geologists call a syncline.

When the Funeral and Black mountains were later uplifted, the basin between them was also brought back up to the surface. Uplift has unleashed the forces of erosion that are now uncovering the geometry of the synclinal fold.

by a fault. From Dante's View, look across Death Valley to the Hanaupah Fan; inscribed across the fan's lower surface is a straight line running north-south. This is a fault scarp with its surface down-dropped on the east side. The fault is most visible when afternoon shadows highlight the fifty foot drop-off.

In our everyday experience, rocks are hard and brittle. But given enough time and pressure, rocks will bend. These upper Paleozoic rocks are at the southern end of the Last Chance Range. Their formations of limestone and dolomite were originally deposited as very flat layers. But this area was badly chewed up by thrust faults and folding, especially during Mesozoic time. The layers, once flat, were tipped every which way, bent into tight folds, and cut by faults. Uplift and subsequent erosion have since been working to expose the rocks.

A fault may easily be overlooked—it might be visible only as the juxtaposition of two different rock types that wouldn't otherwise be found together. Sometimes nothing more than vegetation and color changes will belie the existence of a fault beneath otherwise flat ground. Other geologic phenomena can mimic a fault, such as the old wave-cut shorelines of Lake Manly that we have already discussed. Try not to be fooled; accept only the real thing in your fault-finding.

Once in a while faults are not so subtle. On occasion, they can slap us around. Earthquakes occur when movement suddenly occurs along a fault; then shock waves rock the surface, sometimes miles away from the epicenter of the quake. Californians have long been uneasy bedfellows with the San Andreas Fault; it runs from the Gulf of California, past Los Angeles and through San Francisco, then out to sea in Mendocino County. Death Valley, for all its faults, has not recently experienced nearly as many locally centered earthquakes as the western part of the state. The last significant quake occurred in 1908, apparently centered in the Panamint Mountains. But with so many faults in such a young landscape, it's only a matter of time until another earthquake rolls through, shuffling the deck again.

NORMAL FAULTS typically occur where tension pulls rocks apart, and one block slides down relative to the other block.

Geologists have sub-classified faults into many categories. We will concern ourselves with four types—normal, thrust, low-angle normal, and strike-slip faults. What difference, you might ask, does it make if rocks move up-and-down instead of side-by-side? Lots of difference. These four types of faulting are driven by very different stresses within the earth.

Normal faults. Thrust and strike-slip faults. Tidy categories. As with folds, though, rocks will fault any which way they please. We can certainly find mixed styles of faulting in Death Valley. The Panamint-Hunter Mountain Fault to the west of Death Valley has both a strike-slip and normal sense to its motion—an 'oblique' fault. But for now, let's look at examples within Death Valley that are predominantly one type or another.

A normal fault, like the Hanaupah Fan fault, occurs when rocks on opposite sides of the fault are pulled apart under tension. The plane of the fault tends to be nearly vertical. One block will slide down the fault relative to the other block. The steep west flank of the Black Mountains is made up of many normal faults, all running north-south, with their western sides thrown down. The cumulative throw of these faults accounts for the 9000 foot depth to bedrock beneath Badwater.

A thrust fault, by way of contrast, is caused by the *compression* of rocks. In this case, a block of rock is squeezed by pressure sufficient to fracture it. The rock on one side will tend to slide up over rock on the other side of the fracture. The northern Grapevine Mountains contain a slab of late Precambrian and early Paleozoic rocks, ten miles long and four miles wide; this was squeezed up on top of younger rocks, along the Grapevine Thrust Fault.

THRUST FAULTS occur in a compressional environment, where rocks are squeezed together and once block is forced upward relative to the other.

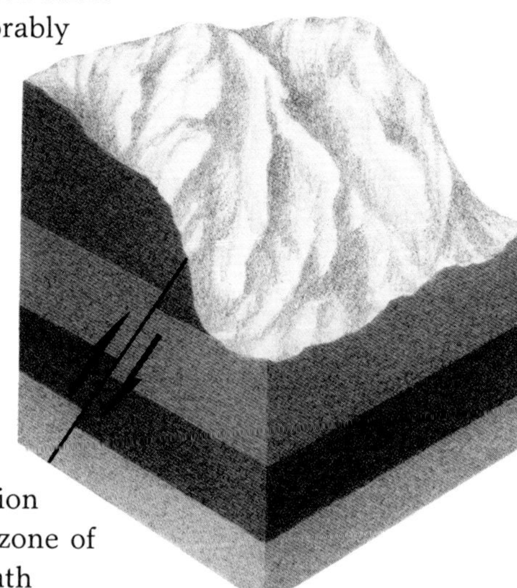

Wait. Don't just read on. Think about that. A slab of rock ten miles by four miles, and thousands of feet thick, once moving inexorably eastward. What awful power utterly beyond the bounds of experience within which we live.

An igneous rock called quartz monzonite intrudes across the Grapevine Thrust Fault, but it was not cut by the fault. The monzonite is radiometrically dated at about 160 million years; therefore movement on the thrust fault had to have ceased by that time. The Grapevine Thrust Fault was one of a series of east-surging thrusts that rolled over Death Valley and the surrounding Basin and Range region during the late Paleozoic and first half of the Mesozoic—the Wheeler Pass, Panamint, Lemoigne, and Racetrack thrust faults.

Once a rock has been fractured, for instance by compression along one of these Mesozoic thrust faults, that fault will be a zone of weakness in the future. In the last fourteen million years, Death Valley has been subject to tension, instead of the compression that dominated during the Mesozoic era. An old thrust fault alongside Bat Mountain appears to have been reactivated, but this time as a normal fault—dropping rock back down the fault that it once rode up.

In the last few years, geologists have surprised themselves by assigning a great deal of importance to low-angle normal faults throughout the Basin and Range. These faults cut great almost-horizontal swaths underneath entire mountain ranges. Fair enough; thrust faults can do this too. These low-angle faults are remarkable, though, because they appear to have occurred in a *tensional* environment. The Basin and Range is being pulled apart along these faults.

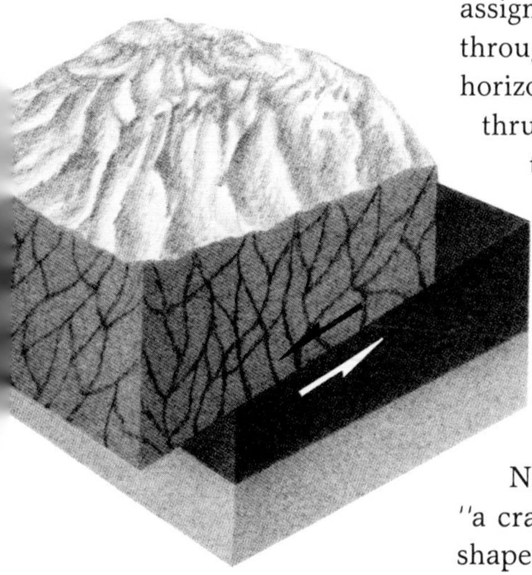

LOW ANGLE NORMAL FAULTS are the intriguing product of a tensional environment. Unlike a typical normal fault, these behave by allowing an upper block to traverse great horizontal distance as it moves down along the lower block.

As early as 1931, Noble was studying the southern Black Mountains around Jubilee Pass. He found evidence of a nearly horizontal fault which he called the Amargosa Thrust Fault. Beneath the fault lay a Precambrian metamorphic gneiss; above it lay a chaotic tangle of fractured rocks.

Noble's scientific vocabulary failed him as he struggled to describe "a crazy-quilt pattern—or more literally a mosaic" of "lozenge-shaped...blocks" with "blunt or rounded ends, whereas others taper to a point; fishlike forms that suggest whales or huge tadpoles or lizards..." Welcome to Death Valley.

The blocks might be a foot long, or half a mile long; all are fractured on every side. A geologic map of this area most closely resembles a shattered windshield. Altogether, this mess is 2000 feet thick. Noble invented a perfect new term for it: the Amargosa Chaos. His astute description of its three phases—Virgin Spring, Jubilee, and Calico—are still the basis for our trying to understand its creation.

But what was once called the "Amargosa Thrust Fault" has since been re-evaluated. Folks now believe that rocks above this fault have most recently been gliding down and to the west, rather than being thrust up to the east. There is growing evidence that many of Death Valley's low-angle normal faults—like the one that underlies the Amargosa Chaos—may originally have been eastward-moving Mesozoic thrust faults that have since been reactivated, now being pulled apart as westward-moving low-angle normal faults.

Geology is an evolving science. The rocks certainly haven't changed all that much during our lifetimes, but our understanding of their origins has grown in leaps and bounds. The outcrops of the Amargosa Chaos that Noble first thought to be evidence for a thrust fault are now interpreted as a low-angle normal fault, caused by tension rather than compression. Geology advances step by step through these sorts of give-and-take growing pains. Only slowly do we gain the insight necessary for a vision that lets us see not only into rocks that are buried thousands of feet below the surface, but also back millions of years through time. What a wonderful lens this science is.

Death Valley contains three "turtlebacks", along the west side of the Black Mountains, at Badwater, Copper Canyon, and Mormon Point. These turtlebacks can be classified as another example of low-angle normal faulting. Their surfaces, pitched at about forty degrees and standing as much as 5000 feet above the valley floor, were scratched and gouged as mountain-sized blocks of rock slid down them to the northwest into Death Valley. Some geologists are convinced that, at some time in the past, blocks of rock slid up instead of down the turtlebacks, but the current consensus is that most recent movements have been down the surfaces.

The underlying geometry of the turtlebacks may have been formed well before Death Valley was created—domed perhaps during Mesozoic or early Cenozoic compression. Alternatively, their surfaces may simply be the more recent result of faulted rocks breaking away and sliding downhill. Where did the rocks originally above the fault go? Some are piled at the bases of the turtlebacks; others were swallowed by the insatiable Death Valley basin.

One last fault type remains—a strike-slip fault. The plane of this fault is roughly vertical, but instead of exhibiting up-and-down motion, rocks move *sideways* along a strike-slip fault. If you stand patiently on one block of rock, and look across a strike-slip fault, rocks on the other side will be moving either to the right or left. One motion is called right-lateral offset, the other obviously left-lateral. The famous San Andreas Fault is a right-lateral strike-slip fault: stand long enough in San Francisco, facing westward, and Los Angeles will eventually sail past you to the right, heading north.

The Furnace Creek Fault Zone is also a right-lateral strike-slip fault. At the north end of this fault zone, about 100 miles from Furnace Creek Ranch, there is good evidence for at least thirty miles of right-lateral offset. Geologists are divided, however, about the total amount of movement that has occurred across the southern part of the fault. One school argues for little more than five miles of offset; the other suggests fifty miles or more.

STRIKE-SLIP FAULTS are vertically oriented faults along which one block slides past another. In a pure strike-slip fault, neither block rises above the other.

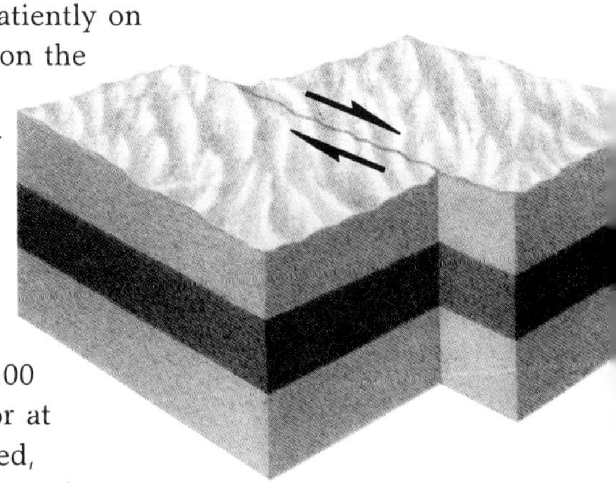

Three *turtleback* surfaces on the western edge of the Black Mountains are all aligned parallel to the direction in which Death Valley is being extended. The turtlebacks appear to be surfaces along which rock has moved to the northwest, down the face of the Black Mountains.

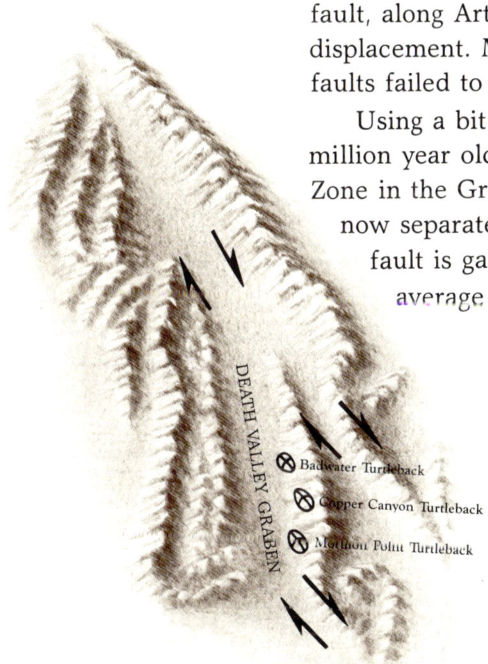

The question remains unanswered.

A second major strike-slip fault is the Southern Death Valley Fault Zone. It has also demonstrated right-lateral offset, totalling twenty miles over the past 10 to 12 million years. A small cinder cone just north of Shoreline Butte is neatly bisected by this fault zone, its southeast corner having been hauled a few hundred yards away, since the time of the crater's extrusion onto the valley floor 690,000 years ago.

How fast are these faults moving? Folks set up precise monitoring stations in 1970, trying to measure offset across some of Death Valley's faults. When remeasured fifteen years later, only one normal fault, along Artist's Drive, had budged: a tenth of an inch of vertical displacement. Measurements across the Furnace Creek and Hanaupah faults failed to show any movement at all.

Using a bit of geologic hindsight, another scientist studied a 7½ million year old granite that was sheared by the Furnace Creek Fault Zone in the Grapevine Mountains. By measuring the distance that now separates the two halves of the granite, he calculated that the fault is galloping along at two-tenths of an inch per year, on average since the granite was intruded.

In 1938, H.D. Curry studied three broadly curved, northwest-dipping surfaces along the western edge of the Black Mountains, descriptively naming them turtlebacks. They have puzzled geologists ever since.

This one, just south of Copper Canyon, stretches from 5000 feet to sea level. The white surface is the lower side of a fault plane. The red rock that is piled up below and to the left of the turtleback, slid as one block down and to the left along the fault. The smooth white surface of the turtleback is gouged with striations, or slickensides, that are all aligned down to the northwest—to the left in this photograph. Erosion has not yet had time to obscure this detailed and delicate evidence of faulting.

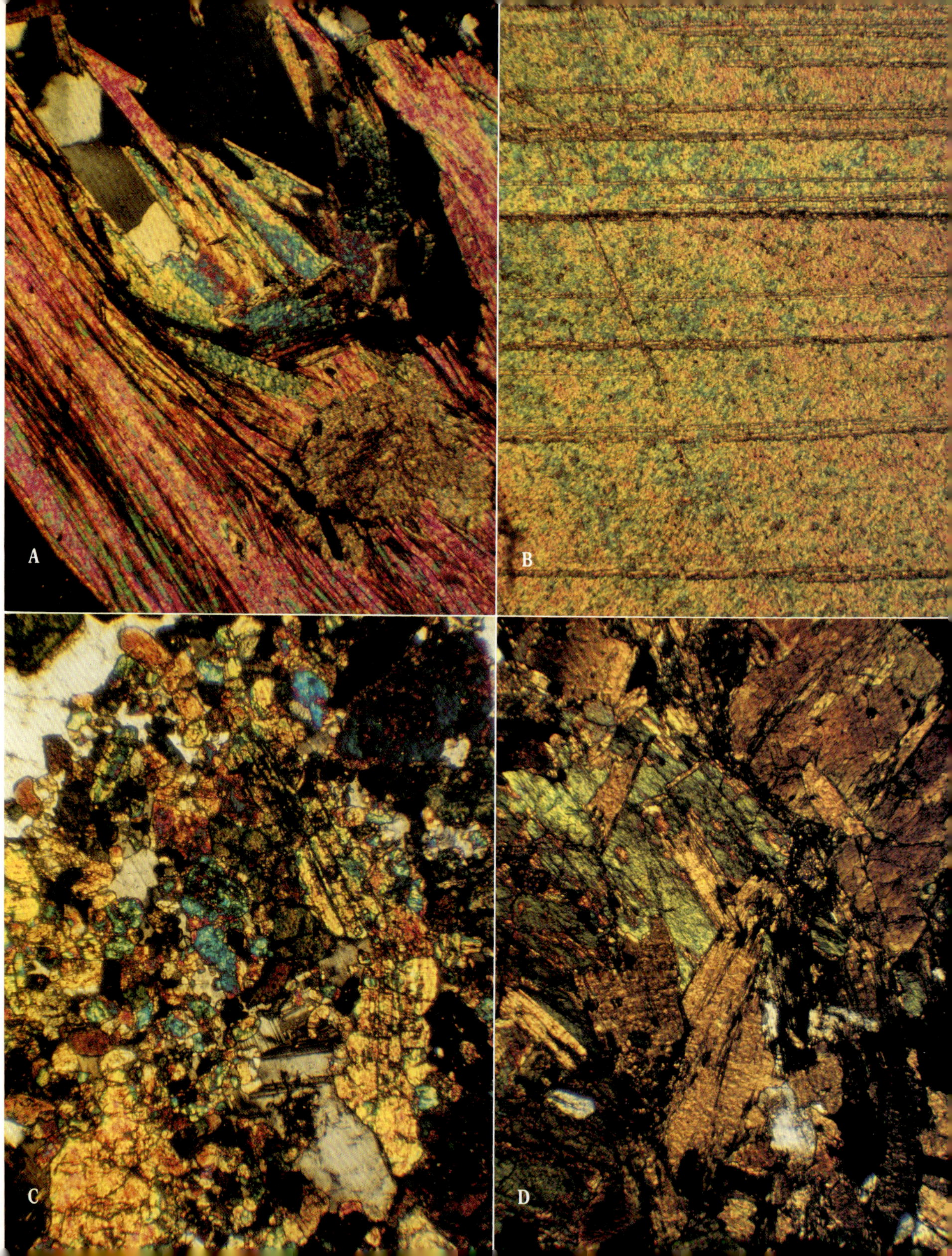
A
B
C
D

PETROLOGY

LOOKING INSIDE ROCKS

I suppose that it is unwise to carry on much longer about geology without talking about rocks. Faults, folds, and gravimeters are all very nice, but ultimately one must return to those cornerstones of geology: rocks. Death Valley is full of them.

I walked up to the Keane Wonder Mine in the Funeral Mountains one cold winter's morning. Had I been looking for gold, I would have been exactly eighty-five years too late: in December 1903 Domingo Etcharren and Jack Keane beat me to it. But I wasn't; I was looking for rocks. I managed to find a few brown ones, black ones, red ones, the usual stuff. But nothing to really get me fired up. Then, just past the mine: a beautiful schist that shimmered, irridescent with muscovite.

A whole new world opens up when we look *inside* a rock. These photomicrographs of thin sections show the interlocking crystals that form A) schist above the Keane Wonder Mine, B) limestone from the Cottonwood Mountains, C) a metasomatized calc-silicate from the Ubehebe mine area, and D) diorite from Smith Mountain.

Another time, high up on the Copper Canyon Turtleback, I traced out a dark grey igneous dike that had been injected up through the Precambrian schist and gneiss of the turtleback. The dike nosed this way and that, seeking out faults, following the weaker planes of foliation, gradually climbing upward through the schist. I lost it where the turtleback had been cleaved smooth by overlying rocks that slid down into Death Valley. I had hoped that it would intrude across the fault, and cut up into the overlying rock. No such luck. Faulting on the turtleback was accomplished long after emplacement of the dike.

Once a friend and I climbed a peak in the Cottonwood Mountains near Dry Bone Canyon. 7342 feet above sea level. 350-million-year-old fossilized corals (of Paleozoic age, more precisely called Devonian) lay scattered just below the peak. These had grown on the bottom of an ocean, to be sealed within the limestone of the Lost Burro Formation, uplifted and only now revealed by erosion. Paleontologists, examining other corals of Devonian age, have found microscopic daily growth rings within annual cycles, thereby unlocking a remarkable secret: a year during the Devonian lasted four hundred days. The earth took just about as long to orbit the sun as it does today, but back then it spun faster on its axis. Science that is able to illuminate such minute and ancient detail is reminiscent of walking from shadow out into hard bright sunlight: it makes me blink.

To be perfectly honest, sorting out all of Death Valley's rock types is not a simple task. It helps to squint, let some of the finer details blur a bit, and initially be satisfied to get the big picture. Sedimentary rocks were deposited in three major phases. Metamorphism and volcanic activity occurred intermittently between and during the longer sedimentary phases.

To be sure, a number of smaller and less obvious phases transpired here, and each phase can be finely divided into many specific strata and formations. But, as one geologic caustic proclaimed, "there is nothing so boring as another man's stratigraphy." So lumpers not splitters we will be.

Let's step back in time: back about 1.4 billion years to the later part of the Precambrian. The west coast of North America then ran through what is now eastern California. Cutting some 140 miles inland (toward Las Vegas) was an embayment with highlands bordering it to the north and south. Lauren Wright and Bennie Troxel are two scientists whose names have become synonymous with Death Valley geology. They refer to this basin as the "Amargosa aulacogen", noting that it looks similar to other 'aulacogens', or embayments, that occur where continents have been incompletely split apart by the tugging of continental drift.

The basin initially experienced slow subsidence. Coarse debris flushed in first from the north, followed by finer silt and sand from both the north and (especially) the south. These sediments were interlayered with carbonates like limestone and dolomite, rocks that form in a warm shallow sea. Algae grew in mats called stromatolites and were fossilized within the carbonates. An igneous rock called diabase squeezed up between layers of the sediments, locally cooking the dolomites and producing talc deposits that have been mined near Saratoga Springs and Warm Springs Canyon. This stack of sediment and diabase, altogether some 3000 feet thick, is now called the Crystal Spring Formation. On top of it were deposited another 1000 feet of dolomite—the Beck Spring Formation. Together, these two formations record a time of relatively slow subsidence within the Amargosa aulacogen.

The floor of the basin began to sink faster, however, as the Kingston Peak Formation was next laid down. Into the basin were dumped sand, silt, and poorly mixed conglomerates in thicknesses

PALEOZOIC	MIS.	Late	Perdido Formation
		E	Tin Mountain Limes
	DEVO.	Late M	Lost Burro Formatio
		E	
	SIL.		Hidden Valley Dolo
	ORDO.	L	Ely Springs Dolomit
		Middle	Eureka Quartzite
		Early	Pogonip Group
	CAMBRIAN	Late	Nopah Formation
		Middle	Bonanza King Form
		Middle / Early	Carrara Formation
		Early	Zabriskie Quartzite
		Early	Wood Canyon Form
PRECAMBRIAN			Stirling Quartzite
			Johnnie Formation

locally totaling 7000 feet. The conglomerates are a wild mixture of boulders, cobbles, and sand, called diamictite. Some geologists have suggested that only glaciers—indiscriminately plowing up all sorts of debris along their path—would be capable of carrying such a hodgepodge of sediments.

Crystal Spring, Beck Spring, and Kingston Peak formations. Together they are known as the Pahrump Group. This group of rocks was tipped by teeterings within the Amargosa aulacogen, and capped by another couple thousand feet of the Noonday Dolomite. Thicknesses within all of these formations often change over relatively short distances, implying that the ground onto which they were deposited was broken by active faults.

Three more formations, the Johnnie, Stirling Quartzite, and Wood Canyon, were deposited after subsidence of the Amargosa aulacogen ceased at the end of the Precambrian era. The Johnnie Formation, unlike the Pahrump Group or the Noonday Dolomite, had a uniform thickness that persisted over huge distances. This uniformity would be typical of deposition throughout the second of our three sedimentary phases. Now, instead of occurring haphazardly in a small active aulacogen, deposition would occur throughout a huge flat shallow sea covering most of western North America.

North America's western shoreline during the latest Precambrian was not far to the east of Death Valley. The sea water was sufficiently stirred up by the crashing of waves and the motion of near-shore currents that it was able to carry sand grains some distance out from shore. This sand was deposited to become the hard white ledges of the Zabriskie Quartzite. But during the early Paleozoic, the shoreline drifted east. Only silt particles could stay suspended in the ocean long enough to reach this area, which by then was a considerable distance out to sea. The silt was dropped here to form the red slopes of the Carrara Formation that is found above Leadfield on the Titus Canyon Drive. Finally, the shoreline shifted so many hundreds of miles to the east that the waters covering what is now Death Valley were clear of most sediments derived from continental exposures to the east.

The remainder of the Paleozoic was dominated by a slow rain of carbonate minerals—either directly from seawater or from the newly evolved shells of sea life. Because the sea was shallow, minor fluctuations in sea level would temporarily bring the shoreline many miles

closer; then mud and some sand would be deposited for a while until the shoreline again receded to the east. The Bonanza King Formation—3600 feet of limestone and dolomite—is the thickest of these Paleozoic carbonates; it forms the walls of the lower part of Titus Canyon Drive in the southern Grapevine Mountains.

The Paleozoic in Death Valley was a sleeper. No earthquakes. A shallow lukewarm sea. No significant topography. Sidewalks rolled up after 6 P.M. To the north and northeast, a flurry of mountain-building activity called the "Antler Orogeny" was taking place, without materially affecting this area. But the pace would have seemed busy enough if you were a trilobite. Sixteen thousand feet of limy mud to squirm around in. Three hundred million years to chase worms and cruise the stromatolites. Not a bad life.

So ends our second sedimentary phase. The Mesozoic era followed, dominated by the thrust faults already described. Intermingled with this faulting (and no doubt related to it) was the intrusion of granite or granite-like igneous rock in pockets scattered throughout the mountains around Death Valley. Hunter Mountain, composed of 165-million-year-old granite, was one of the larger pockets or plutons of granite emplaced in this vicinity. To the west, the granites that now form the Sierra Nevada were being intruded far underground; later they would be uplifted to become the mountains we know today.

Death Valley has dropped down the west side of the Black Mountains along a normal fault, in which movement is straight down the fault plane. Artists Drive here winds along a straight scarp – the fault's break in the land surface that has not been significantly erased by erosion.

The Black Mountains, on the right side of this scarp, are here composed of the Artists Drive Formation laid down six to fourteen million years ago. The alluvial fan on the left side of the fault leads down to the Death Valley playa at the Devil's Golf Course. The fault itself draws an almost straight line across the surface here, suggesting that fault plane is nearly vertical. The faults that separate Death Valley from the Black Mountains are thought to be steep near the surface but with depth flatten toward the horizontal.

The granites at Skidoo have been dated at 67 to 87 million years, the very end of the Mesozoic. Two prospectors stumbled across the white quartz ledges of the Skidoo Granite in 1906; legend has it that the two men were lost in a fog at the time. True to their prospecting heritage, they were happy to sell their claims to a mine promoter for $30,000 apiece and move on. The promoter, a Bob Montgomery of Rhyolite, spent an additional quarter million dollars on a pipeline that brought much needed water from the flanks of Telescope Peak to the boom town that was developing around the claims. The pipeline was eight inches in diameter and twenty-three miles long. Given the popularity of the saying "23-Skidoo," that pipeline may have been good enough an excuse for the town's name. Skidoo eventually produced $1,344,500 worth of gold, making it one of the few truly profitable gold mines developed at Death Valley.

The thrust faults and granite intrusions in Death Valley are solid evidence that the Mesozoic was a time of tremendous heat and pressure around Death Valley—exactly the right setting for the creation of metamorphic rock. The Panamint and Funeral mountains both contain large bodies of schist and gneiss, metamorphosed not only from older Precambrian bedrock but also from some of the younger Precambrian sedimentary units like the Pahrump Group and Noonday Dolomite. Mineral assemblages in the Funerals suggest that those mountains experienced greater heat and pressure than did the Panamint Mountains.

A layer (actually a series of layers) of volcanic ash buried most of Nevada and the Death Valley area beginning 27 million years ago. The layers piled up as deep as 1200 feet across the Grapevine Mountains. Many thousands of cubic miles of the ash exploded from volcanic centers near the present Nevada Test Site, southeast of Beatty. The ash is of rhyolitic composition, chemically equivalent to granite; had it remained underground, it might have become the core of a Sierra Nevada-like mountain range. Instead, it was blasted out into the atmosphere, and settled to erase all vestiges of earlier topography. Looking east from Death Valley twenty million years ago, you would have seen virtually a single seared plain stretching hundreds of miles across Nevada.

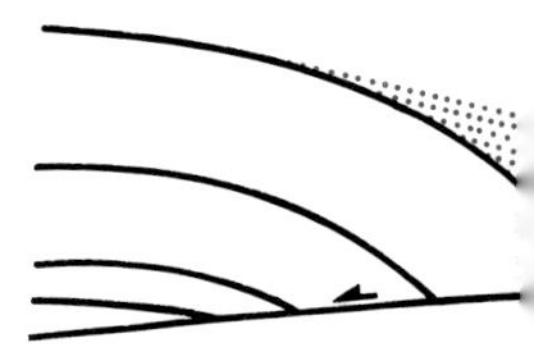

It is hard to go far back in time, and keep track of so many oceans, thrust faults, and intrusions. It's difficult to imagine Death Valley in any form but the mountain-and-basin arrangement we see today. At different times in the past, though, this area has looked a lot like the Canadian Rockies, Bahamas, Sahara desert, Mt. St. Helens, or the south side of Chicago. But those rhyolitic ash layers swept the slate clean, like a child smoothing off the ground before drawing a map with a stick. Now we can watch the modern Death Valley rise like the phoenix fresh from its own ashes. And so begins our final phase of building the rocks that today comprise Death Valley.

Strike-slip movement began along the Furnace Creek fault zone fourteen million years ago. As the fault began to move, a basin was formed along what is now Furnace Creek. And nature, abhorring a vacuum or even a low spot, immediately started to fill the basin with sediments, mostly derived from the south.

The Black Mountains had just begun to rise as an identifiable range; into this basin, they shed conglomerates, fragments of Paleozoic sedimentary formations like the Bonanza King, as well as granite fragments from the recently emplaced Smith Mountain pluton. This material, when stirred together with a few ash deposits, would become the 4000-foot-thick Artist Drive Formation. Radiometric dating of the ash confirms the age of this formation as 14 to 6 million years old. It is easily spotted today as the red, green, and turquoise colored cliffs and slopes along Artist Drive.

The Furnace Creek Formation was next deposited in the river and lake sediments intermixed with more volcanic material, altogether 7000 feet thick. Boron was leached from the surrounding volcanic terrain, and washed into the basin; this was concentrated into the colemanite borate deposits that have since been mined around Ryan. Last to arrive in the Furnace Creek Basin was the Funeral Formation—2000 feet of (you guessed it) sand, mud, conglomerates, and volcanics. A basalt flow of the Funeral Formation in the Greenwater Mountains has been dated at four million years. Contemporaneous with deposition of the Furnace Creek and Funeral formations was that of the Copper Canyon Formation in a smaller basin to the south. Fossils of camels, mastodons, sabertoothed cats, and horses have been found within one or the other of these three rocks.

The Nova Formation has been deposited in a trough formed by extension between the Panamint and Hunter Mountains. It appears that blocks of the Panamint Mountains have successively faulted into the basin as extension occurred.

Other rocks have made their appearance in and around Death Valley since the deposition that occurred within the Furnace Creek basin. Most are localized within smaller areas. The Nova Formation, for instance, is a 6000-foot stack of sedimentary and volcanic rocks lying northwest of the Panamint Mountains. This formation was deposited 3½ to 5 million years ago in a trough created by movement on faults that are now found near Tucki Mountain.

Ubehebe Crater was formed only a few thousand years ago when magma rose close enough to the surface to encounter ground water. The water was flashed to steam, and the sudden extreme pressure caused the crater to violently explode, creating what is known as a maar crater. A number of similar smaller craters around Ubehebe suggest that these explosions occurred a number of times.

What was once one hill is now two. This cinder cone erupted onto the surface of Death Valley 690,000 years ago. The cone is split by a fault that is part of the Southern Death Valley Fault Zone. Movement along the fault has cut the cinder cone in two and carried the smaller part a few hundred feet to the southeast. Knowing the age of the cinder cone and the amount of offset, we have a rough idea of the rate of the fault's movement over the past 700,000 years.

Movement along the Southern Death Valley Fault Zone is primarily side-by-side; therefore it is called a strike-slip fault. By examining the fault in other locations, geologists have concluded that it started to move about 10 to 12 million years ago. Volcanic eruptions have occurred throughout Death Valley. Often they reach the surface along a fault. Faults present a weakness in the earth's crust up which magma is wont to move. Shoreline Butte lies just south of this cinder cone and along this same fault; it erupted 1 1/2 million years ago.

EXTENSION

PULLING THE EARTH APART

Even though it was the Fourth of July, Pete Aguereberry was not inclined to race. He eyed the two kids who were cantering around in circles, warming up in running shoes and funny little shorts. It was a good bet that one of them would win this two-hundred-and-twenty-yard foot race. They were engineering students at Berkeley, spending the summer of 1914 working in the mines of Skidoo. Rumor had it that both kids were on the school track team. And Pete was just a busted up 39-year-old hardrock miner.

Nine years earlier, Pete Aguereberry had started the Harrisburg gold rush when he found a paying vein in the northern Panamints. Since then, he had divided his time about equally between mining and beating away claim jumpers. Pete, a Basque emigrant, had learned about herding sheep, driving freight wagons, and then mining—all the hard way: he would roll up his sleeves and jump in. But he also knew something about running races. He had won this race every year that it had been held—except once when somebody tripped him.

Well, it *was* July Fourth and the boom-town folks of Skidoo weren't about to let the depredations of old age get in the way of a good race. So Pete gave in. He was bean-pole tall, dressed in coveralls. In deference to the race, he had kicked off his boots. The two kids were in a racer's crouch, waiting for the start. The gun sounded and the kids leapt off the line, leaving Pete in the dust.

But there's more to a race than just the starting gun. Pete gathered momentum, pounding away in his stocking feet, coveralls flapping. He passed first one, then the other kid just as he crossed the finish line. Pete had won again. The crowd, of course, went nuts.

There have been more things running around Death Valley than just miners. Entire mountains, for instance. Death Valley is bracketed by two major fault zones—the Furnace Creek fault zone on the north, and the Southern Death Valley fault zone on the south. Both are strike-slip with right-lateral offset. The Furnace Creek fault zone appears to have started moving 14 million years ago; the Southern Death Valley fault perhaps a couple million years later.

Black Mountains at Copper Canyon.

Geologists have been bickering for years about the total amount of movement on these faults. Wright and Troxel suggest a conservative three to five miles, based on sedimentation similarities they have observed in exposures of the Pahrump Group on either side of the Furnace Creek fault zone. Jack Stewart and Brian Wernicke are long-time standard-bearers of another camp of geologists, who argue for some fifty miles of movement across this fault.

Both camps agree that the faults at the north and south ends of Death Valley are not connected. And both are convinced that, regardless of the total amount of offset assigned to these two faults, their combined effect has been to place a tremendous oblique tension on the land lying between them. This tension first opened the Furnace Creek Basin. Death Valley itself opened later in stages; the basin at Badwater was one of the last, pulled apart about 4 million years ago.

The Black and Panamint mountains stare at each other across Death Valley, engaged in a tug-of-war of geologic proportions. And Death Valley is caught in the middle: stretched and dropped below sea level. What are the rules of this game?

Jack Stewart observed that the Panamint Mountains contain a thick pile of late Precambrian and Paleozoic sedimentary rocks. The Black Mountains, however, are basically older metamorphic Precambrian rocks with a cover of Cenozoic sedimentary and volcanic rocks—none of the Precambrian and Paleozoic strata found in the Panamints. Stewart proposed that the Panamint Mountains at one time roosted squarely on top of the Black Mountains.

As movement began on the Furnace Creek and Southern Death Valley fault zones, tension grew to such proportions that the Panamint Mountains slid west down a low-angle normal fault, off the back of the Black Mountains. Stewart suggests that the three turtlebacks that now stand above Death Valley may well have been surfaces along which some of this low-angle normal faulting took place. This movement would had to have been accomplished as of 6 million years ago, because the Furnace Creek and Funeral formations of that age are now found atop the Black Mountains. Death Valley, according to this proposal, is the pull-apart basin left in the wake of the departing Panamints.

To appreciate the audacity of this proposal, you really do need to dangle your feet over the edge at Dante's View in the Black Mountains, and look twenty miles across to the Panamints. That's a lot of mountain

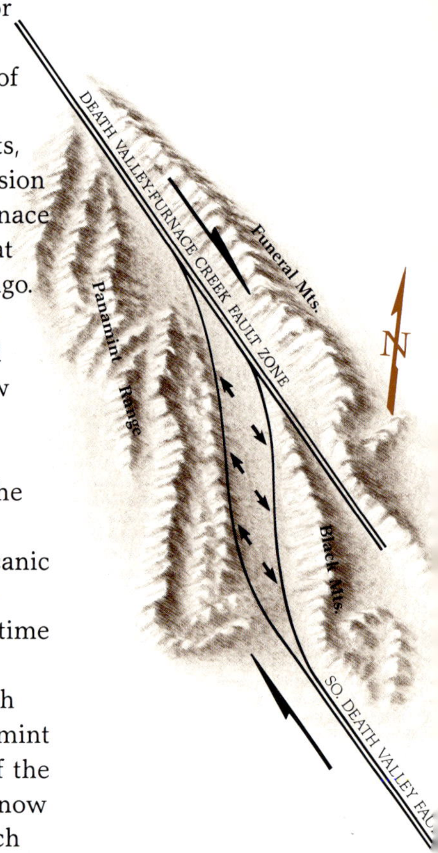

over there to drag around—fifty miles long, looming two miles up in the sky. We're not talking Twenty Mule Team cargoes here. We're slinging whole mountains around like they were chips in a checkers game. This is not that subtle science that makes one blink; this is sledge-hammer-between-the-eyes geology.

All the same, Stewart just may be right. Others of his camp have provided impressive studies correlating offset folds and thrust faults, tending to corroborate this seemingly outrageous proposal.

But Wright, Troxel, and other geologists have provided equally impressive—albeit contradictory—studies of faulting along the western face of the Black Mountains. Collectively they have logged decades in the field, studying the composition of the Furnace Creek basin deposits. They argue, quite convincingly, that the Panamint and Black Mountains started out side-by-side, and have since been pulled apart along normal faults. They suggest that these faults are initially steep but curve with depth to become low-angle faults. The mountain blocks rotate as they slip down these curved faults, creating the tilted mountains with their asymmetric basins. This pulling apart, they argue, has been enough to create Death Valley.

Death Valley lies at the intersection of a number of major fault lines. Bound by the parallel Death Valley-Furnace Creek and the Southern Death Valley Fault Zones, Death Valley itself is steadily being pulled apart.

Who are we to believe? A bit of both perhaps. The science of geology thrives on this sort of intellectual sparring. Out of these disagreements grows a dynamic understanding of the earth and her evolution. The soil of Death Valley—though scorched and salt-ridden—has been very fertile for geologic thought. Death Valley, at the southwestern corner of the Basin and Range Province, has provided a new model for extension of the earth's crust.

Scientists now think that the Basin and Range has been pulled apart as much as 150 miles, along a line drawn through Death Valley between Lake Mead and the southern Sierra Nevada. The Sierras, they say, have been migrating west, away from the rest of North America for the past 15 million years. At first, extension occurred at a rate of one inch per year, but slowed to a third of that over the last 5 million years. Pete Aguereberry might have outrun them for a while. But the earth has stamina: 15 million years' worth.

Extension of this magnitude must certainly be driven by the whims of drifting continents. Also known as plate tectonics, this theory choreographs the dance of oceans and continents around each other. The basic building blocks of plate tectonics are pieces of crust: ocean crust and continental crust. The crust beneath the ocean tends to be thin—perhaps five miles, while crust beneath continents is usually at least twenty miles thick. Geologists have outlined about twenty major plates throughout the world—some are pieces of oceanic crust, others continental. Both types float on top of the earth's next deeper layer, the viscous mantle. The continental plates are, on average, made of lighter rocks than the oceanic plates; consequently the continental plates float a bit higher upon the mantle.

To apply the theory locally in its simplest form: the North American continental plate was overriding the Pacific plate until sometime in the late Cenozoic. Then interaction between the two plates shifted, and they began to slide past one another, with the Pacific moving north relative to North America.

The San Andreas Fault is the boundary between these two plates, and the abrasive energy of their relative motion is focused on the fault. The Basin and Range Province, however, is wholly contained within the North American plate, hundreds of miles inland from the plate boundary. Even so, it has been stretched by the same forces that are driving the two plates.

Instead of tearing along one large fault like the San Andreas, the Province has been stretched along many smaller strike-slip and normal faults. Much of the Basin and Range's extension is occurring along low-angle normal faults. East of Death Valley, some areas of the Basin and Range have been stretched like taffy, rather than breaking along faults. One result of this overall stretching is that the Basin and Range crust is anomalously thin—fifteen miles. Continental crust on either side of the province is much thicker: twenty-five miles beneath the Colorado Plateau, and thirty miles thick beneath the Sierra Nevada.

Geologists have come a long way since Charles Breyfogle stumbled into Las Vegas Springs, incomprehensibly babbling about his Lost Ledge. Breyfogle would likely have thought we were crazy to be talking about this theory of plate tectonics that has whole continents pirouetting around the globe. Standish Rood accompanied Breyfogle back into Death Valley in 1866, looking for that old Lost Ledge. Rood described his impressions six years later:

> It is sad to see a man of his age wearing the few remaining years of his life out in wandering about the desert valleys and desolate mountains of

> this godforsaken, sterile country, mounted on an old Piute pony, following a still poorer one ladened with all his worldly goods, consisting of a chunk of old bacon, a sack of flour, a battered iron coffee pot, and a greasy frying pan...

Our understanding of Death Valley geology has steadily grown from its early days of hit-or-miss prospecting to today's geophysical wizardry. Concepts like low-angle normal faulting and extensional tectonics are still being worked out in this great natural laboratory. In 1988 Brian Wernicke concluded a milestone paper on Basin and Range extension by reaffirming that geologists must always return to the field, get down on their hands and knees, and truly look at the rocks. The theories may evolve, the laboratories back home might be fancier, but the things that are most important have remained the same. What is essential is that we have been out here and have seen the rocks.

Geology, like prospecting, in some ways will always be a private affair. I hiked to the top of the Funeral Mountains one day, past a crumbling mill and a couple of long abandoned mine shafts. The wind was blowing and I thought I could hear the sad slow clanking of those pots and pans tied to Breyfogle's pony, coming behind me up the hill. No other place on earth, I thought, could create and then conceal a Lost Ledge so well. Within Death Valley, rocks are deceptively barren, tantalizing in their degree of exposure. The geology—and prospecting—should have been so easy. But each new strike only led to one more Lost Ledge; and nowadays each new geologic theory only leads to more questions. Geology is not a static science. Plate tectonics has so completely rewritten our appreciation of the earth's evolution that it is no longer sufficient to simply *describe* places like Death Valley—now we strive to *understand* them.

Atop the Funerals, I surveyed the world that had been special to Breyfolge and, in turn, has been special to each one of us for a century and a half. Below lay Furnace Creek, where Rood had regarded Breyfogle with feelings that bridged pity and contempt. The Panamints, where Pete Aguereberry had persistently mined and occasionally raced. The Stovepipe sand dunes, where only the day before I had rejoiced in the beauty born of wind and time. I watched it all, listening to the sound of pots and pans moving softly away on the wind.

BIBLIOGRAPHY

Cemen, I., L.A. Wright, R.R. Drake, and F.C. Johnson. Cenozoic Sedimentation and Sequence of Deformational Events at the Southeastern End of the Furnace Creek Strike-Slip Fault Zone, Death Valley Region, California. *Society of Economic Paleontologists and Mineralogists Special Publ. No. 37.* 1985.

Fiero, Bill. *Geology of the Great Basin.* University of Nevada Press. 1986.

Greene, L.W. and J.A. Latschar. *Historic Resource Study: A History of Mining in Death Valley National Monument.* National Park Service; Denver, Colorado. 1981.

Hildreth, Wes. *Death Valley Geology.* Death Valley Natural History Association. 1976.

Hunt, Charles. *Death Valley.* University of California Press. 1975.

Lingenfelter, R.E. and R.A. Dwyer. *Death Valley Lore.* University of Nevada Press. 1988.

Lingenfelter, Richard. *Death Valley and the Amargosa.* University of California Press. 1986.

Manly Beacon from Zabriskie Point.

Stewart, John H. Extensional Tectonics in the Death Valley Area, California: Transport of the Panamint Range Structural Block 80 Km Northwestward. *Geology,* v. 11. p. 153. 1983.

Troxel, B.W. and L.A. Wright, editors. *Geologic Features, Death Valley, California.* California Division of Mines and Geology. 1976.

Wernicke, B., G.J. Axen, and J.K. Snow. Basin and Range Extensional Tectonics at the Latitude of Las Vegas, Nevada; *Geological Society of America Bulletin,* v. 100, p. 1738. 1988.

ACKNOWLEDGMENTS

An introductory geology book will always owe its greatest debt to the geologists who have done the years of field and laboratory work that necessarily precede this sort of compilation. Mel Essington, Death Valley's mining engineer, has been helpful all along. Ibrahim Cemen, Kent Snow, Ronald Dorn, Bennie Troxel, Lauren Wright, and Isaac Winograd were kind enough to suggest and send some of their current contributions to the geologic literature. Ray Lund of Quality Thin Section in Tucson generously volunteered to prepare rocks for photomicrography. Ed Peacock was a great help with aerial photographs. My wife, Rose Houk, was supportive not only on occasion in the field, but throughout the whole process of creating this book. Esy Fields, of the Death Valley Natural History Association, has been invaluable; without her encouragement and enthusiasm, I am sure that this book would never have seen the light of print.